591.95

UNSWORTH ELEMENTARY SCHOOL
5685 Unsworth Road South
Chilliwack, BC V2R 4P5

INTO Wild India 2

BLACKBIRCH PRESS
An imprint of Thomson Gale, a part of The Thomson Corporation

THOMSON
GALE

Detroit • New York • San Francisco • San Diego • New Haven, Conn. • Waterville, Maine • London • Munich

© 2005 Thomson Gale, a part of The Thomson Corporation.

Thomson and Star Logo are trademarks and Gale and Blackbirch Press are registered trademarks used herein under license.

For more information, contact
Blackbirch Press
27500 Drake Rd.
Farmington Hills, MI 48331-3535
Or you can visit our Internet site at http://www.gale.com

ALL RIGHTS RESERVED
No part of this work covered by the copyright hereon may be reproduced or used in any form or by any means—graphic, electronic, or mechanical, including photocopying, recording, taping, Web distribution or information storage retrieval systems—without the written permission of the publisher.

Every effort has been made to trace the owners of copyrighted material.

Photo credits: cover, all pages © Discovery Communications, Inc. except for pages 6-7, 23 © Photos.com; pages 11, 14, 37 © Corel Corporation; pages 18-19, 43 © PhotoDisc

Discovery Communications, Discovery Communications logo, TLC (The Learning Channel), TLC (The Learning Channel) logo, Animal Planet, and the Animal Planet logo are trademarks of Discovery Communications Inc., used under license.

LIBRARY OF CONGRESS CATALOGING-IN-PUBLICATION DATA

Into wild India 2 / Marla Ryan, book editor.
    p. cm. — (Jeff Corwin experience)
  Includes bibliographical references and index.
  ISBN 1-4103-0233-4 (hardback : alk. paper) — ISBN 1-4103-0234-2 (pbk. : alk. paper)
 1. Lions—India—Gir Forest National Park—Juvenile literature. 2. Tigers—India—Gir Forest National Park—Juvenile literature. 3. Wildlife conservation—India—Gir Forest National Park—Juvenile literature. 4. Zoology—India—Juvenile literature. 5. Corwin, Jeff—Travel—India—Juvenile literature. I. Ryan, Marla Felkins. II. Corwin, Jeff. III. Series.

QL737.C23I59 2004
591.954—dc22
                                            2004004484

Printed in the United States
10 9 8 7 6 5 4 3 2 1

Ever since I was a kid, I dreamed about traveling around the world, visiting exotic places, and seeing all kinds of incredible animals. And now, guess what? That's exactly what I get to do!

Yes, I am incredibly lucky. But, you don't have to have your own television show on Animal Planet to go off and explore the natural world around you. I mean, I travel to Madagascar and the Amazon and all kinds of really cool places—but I don't need to go that far to see amazing wildlife up close. In fact, I can find thousands of incredible critters right here, in my own backyard—or in my neighbor's yard (he does get kind of upset when he finds me crawling around in the bushes, though). The point is, no matter where you are, there's fantastic stuff to see in nature. All you have to do is look.

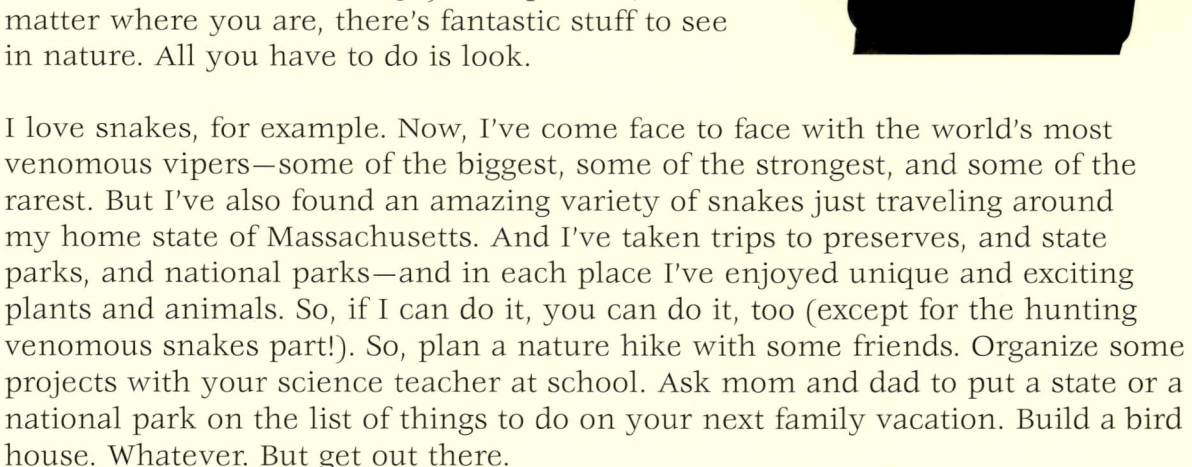

I love snakes, for example. Now, I've come face to face with the world's most venomous vipers—some of the biggest, some of the strongest, and some of the rarest. But I've also found an amazing variety of snakes just traveling around my home state of Massachusetts. And I've taken trips to preserves, and state parks, and national parks—and in each place I've enjoyed unique and exciting plants and animals. So, if I can do it, you can do it, too (except for the hunting venomous snakes part!). So, plan a nature hike with some friends. Organize some projects with your science teacher at school. Ask mom and dad to put a state or a national park on the list of things to do on your next family vacation. Build a bird house. Whatever. But get out there.

As you read through these pages and look at the photos, you'll probably see how jazzed I get when I come face to face with beautiful animals. That's good. I want you to feel that excitement. And I want you to remember that—even if you don't have your own TV show—you can still experience the awesome beauty of nature almost anywhere you go—any day of the week. I only hope that I can help bring that awesome power and beauty a little closer to you. Enjoy!

Best Wishes!
Jeff

INTO Wild India 2

NORTH AMERICA

SOUTH AMERICA

Atlantic Ocean

Pacific Ocean

EUROPE

AFRICA

ASIA

Indian Ocean

AUSTRALIA

Pacific Ocean

It is as exotic as it is mysterious. It is a land of a billion people and some of the most pressured wildlife on our planet. Both human and animal are fighting for survival. Join me as we search for two creatures caught in the middle of this survival challenge. They are two of the world's most famous predators, the tiger and the lion.

I'm Jeff Corwin.
Welcome to India.

I've already made a tractor full of friends!

Getting around India can be an experience in itself. I've hitched a ride on a tractor to our first stop. No air conditioning, airbags, or seatbelts — but it sure has a lot of fun people. That's what's great about traveling in India, it's always an easy place to make friends.

We've just said good-bye to the tractor, and now look at this. It's an Indian cobra— an extremely venomous

as well as the famous Indian cobra.

6

India is home to the famous Taj Mahal...

I'll get you out of there.

My hands are trembling...I just love this snake!

serpent. I really have to concentrate because not only is this beautiful snake venomous, he's in a really tough position, in a gnarly thorn bush.

To catch him I have to be very gentle, very careful—because I don't want to pin him to the point that I hurt him. And I have to be quick because he's lightning fast. Here we go.

Ooh, my hands are trembling. My hands always tremble when I have a snake in them. It's because I just love them.

Just look at that face.

Aren't these cobras beautiful?

These are beautiful snakes, these Indian cobras. This is what I love about exploring the subcontinent of India—you can often find snakes living close to people. Here, we've just stepped off the road. We can still hear

9

Humans and cobras live side by side here.

There you go. You're free again.

the traffic. And here, moving about in this little patch of wilderness, is this amazing serpent.

Cobras are fairly common throughout India. They do well in human habitat, where they'll feast on rats and vermin—creatures that are actually pests. To a farmer, this snake performs a very valuable service.

These animals are not only recognized as being part of the wilderness of India, but they're also recognized as being part of the culture. So let me release this wonderful serpent.

This wonderful serpent is an important part of the culture.

This banyan tree is unbelievable!

Can't you see it crawling with these roots?

Check out this tree—it is unbelievable. This is a banyan tree, one of the oldest and largest banyan trees in India. You can see why many people of this part of the world think of the banyan tree as the walking or crawling tree. The tree grows by sending out prop roots that develop into trunks, providing support as well as nourishment for the tree. Over centuries a tree like this one actually moves through its habitat, by sending out new roots and forming new trunks.

I hope I blend in with these Hanuman langurs.

No, that's not me. It's a monkey.

Buddha spent six years meditating beneath a banyan tree just like this, which contributed to the great religion of Buddhism. But this tree is not only important to humans. It's also important to wildlife. In fact, just to my right are a couple of really nice animals, monkeys.

Give me a cobra, and I have no problem. Surround me with a bunch of primates, and I feel a little iffy. But let's see how I do with these guys. These are Hanuman or black-faced langurs. They make their homes in everything from rain forests to dry forests, and they're pretty social. Hopefully, they're looking at me as another black-faced langur just hanging around. Their diet is almost 100 percent vegetarian. They'll eat an insect or two, and maybe sneak in a bird's egg, but they prefer fruits, fungus, and leaves.

If you look him in the eye, he'll take it as a challenge.

One monkey is actually reaching into my shoes. I don't know if that's a compliment or an insult. Maybe he's grooming me.

When I look at this creature I have to be careful not to take it on eye to eye. If you look many nonhuman primates in the eye, they take it as a threat or a challenge. So you just want to keep looking up and down, up and down.

If you were here now, you'd know that these animals can be pretty vocal. Ah, ah, ah, ah, ah— that's a call of fear or disturbance. I don't want to say it too loud because I don't want to psych out these animals. If you hear woo, woo, woo, that's a happy call.

Okay, everybody. Can I get a woo?

Believe me. You don't want my shoe.

Can I get a woo woo?

India was once home to thousands of Bengal tigers.

But now there aren't many left.

SAVE THE TIGER
OUR NATIONAL ANIMAL
80 YEARS AGO INDIA ROARED WITH
40,000 TIGERS
NOW ONLY
1800

Centuries ago, the tiger truly did rule this land. But unfortunately, because of years of hunting and loss of habitat, the population of the Bengal tiger has been decimated. Even though the mighty tiger has been officially protected for generations, there will never be as many tigers as there were just twenty-five years ago. Today, if you want to see a Bengal tiger in the wild, the best place to do it is Ranthambhor National Park.

16

Fateh Singh is a master tiger tracker.

He's made Ranthambhor National Park a safe home for them.

    This is Fateh Singh, master tiger tracker. As the park's director for thirty years, he was responsible for literally saving the tigers here from extinction. He helped secure and police the park's borders, keeping the tigers in and the poachers out. Today, Ranthambhor is a success story, mostly because of Fateh Singh. He is a legend among conservationists worldwide. And he's going to help us find tigers.

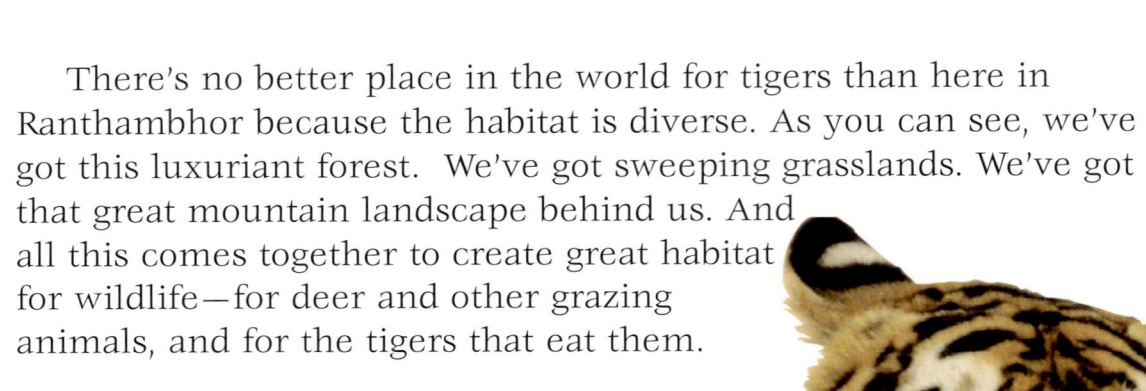

There's no better place in the world for tigers than here in Ranthambhor because the habitat is diverse. As you can see, we've got this luxuriant forest. We've got sweeping grasslands. We've got that great mountain landscape behind us. And all this comes together to create great habitat for wildlife—for deer and other grazing animals, and for the tigers that eat them.

These spotted deer are the tigers' main prey.

Here we have a herd of cheetal or spotted deer. There are a lot of these animals in the park, and they're a main prey item in this ecosystem. What's interesting about these deer is that the herds are, for the most part, matriarchal, female led. The males, or bucks, compete with each other to be selected by the females, to be entered into the harem. And the harem is not controlled by the male, but by the females.

Check this out. This is a tiger track. Fateh says it was made by a medium-size tiger, maybe a four-year-old cat that's not completely mature.

Check out this tiger track.

The Bengal tiger is India's greatest predator.

19

Whoa—look at this. It's tiger tatti. Tatti is the informal word for scat in Hindi. And this is fresh. It stinks. Ugh! But by breaking it up, we can find out what this tiger had for dinner.

What did this tiger have for dinner?

There is fur in the scat, and it looks like the fur of a sambar, the largest Indian deer. Everything else has been digested—bone and flesh. But the fur actually serves as roughage. So while you're getting your bran to keep the pipes flowing, the tiger is using the fur of its prey to keep his intestines in proper working order.

Looks like sambar fur.

And as you can see, that's a nice clump of scat, fresh enough that we know we're getting close. Keep in mind that this huge place—five hundred-plus square miles of protected habitat—has only thirty-eight tigers.

20

Fresh tiger scat really stinks.

There's movement at the riverbank.

There she is! A beautiful Bengal tiger.

Fateh and I have been tracking this tiger since the morning, and now the sun is setting. We found tracks. We found some scat. And finally, our persistence is paying off. We've seen movement in the bushes. An animal has been working its way down this riverbank, and it is the beast we've come to find here.

Look at this—a beautiful *panthera tigris tigris*, the Bengal tiger. It's checking out its terrain. This is India's greatest predator. It's a fantastic creature when it comes to taking its prey, so powerful and so huge. Its paw is twice the size of my hand.

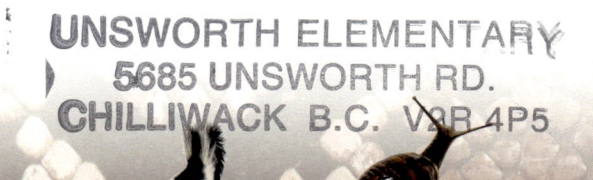

UNSWORTH ELEMENTARY
5685 UNSWORTH RD.
CHILLIWACK B.C. V2R 4P5

21

"Her belly's kind of rounded."

"She's going to be a mother soon."

Fateh tells me that this is a female, and she's almost ready to give birth. A female like this can give birth to up to six cubs, but usually only two will survive. Why? You have competition between cubs. You also have competition between adult tigers.

Bear in mind that female tigers have to sustain themselves throughout their pregnancies without help. They don't work in a pack. A pregnant lion can rely on another lioness to catch prey, and then move in and take part of the kill. It doesn't work that way with the tiger, and that's why a tiger's pregnancy doesn't last very long—a little over three months.

> Tiger cubs are totally dependent on their mother.

The cubs are born helpless, completely dependent upon mother, blind and tiny. Chances are some will be born dead. And among those that are born alive—despite the fact they have this great, powerful mother who puts her life on the line to defend her young—there will be failures. In the end, you'll usually have two that survive.

A sambar has wandered a little too close.

Mom just missed a meal.

Oh, she hears something. It's a deer, a sambar. The tiger is hiding behind a rock, and the deer is moving right in front of her. I cannot believe this....

Just at the last minute, the sambar smells the tiger and takes off. That's one lucky sambar. Now it's warning all the sambar deer. And we're losing the light. But there's more to discover tomorrow.

Where did this guy get his license?

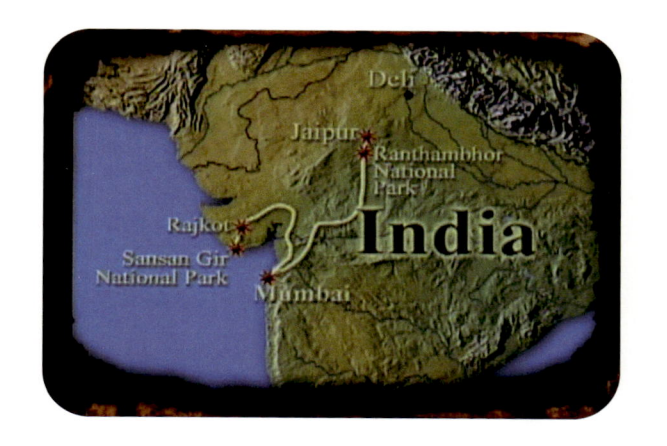

Put a cobra in my hands, no problem. Put me on the teeming streets of Jaipur, and my heart starts racing. When you're moving through the streets in a pedicab, you put your life in the driver's hands. I hope he doesn't get me killed.

This will be a 15-hour travel day. It began with a bus ride and then this pedicab. From Jaipur, we'll fly to Mumbai, which was formally known as Bombay, and then on to Rajkot. Then we'll travel for another four hours by bus to reach our destination—Sansan Gir National Park, where I hope we'll come face to face with India's other great feline predator, the Asiatic lion.

We've got lions to see!

It's hot and dry here in Sansan Gir...

the only place to find Asiatic lions.

Of course, you'll remember nothing of this. When I snap my fingers, you'll wake up in Sansan Gir.

Sansan Gir is located in an arid region in western India. Gir is unique because it is only in this national park that we can find the last remaining population of the Asiatic lion. But many of India's indigenous animals can also be found here.

26

We've spotted a lot of scat; my guess is that it's cheetal. And we passed a hole that could be the entrance to a wild boar's den. And now, right in front of us, is one of my favorite Indian lizards.

Check it out. It's a brilliant, wonderful lizard. This is the *Uromastyx*, also called the spiny-tailed lizard. This animal has dug itself a burrow in a hillside, where it has an excellent view of any oncoming predators. It doesn't feel threatened by me because it knows that if I do something tricky, it can just turn around, bottom over tea kettle, and tumble down into its tunnel. The tunnel may go two or three feet into the earth.

So how are we going to extract this creature? I'm going to move in very slowly. These lizards are not known for biting.

There's a spiny tailed lizard down there.

Its tunnel goes down a few feet.

There he is!

He doesn't look too happy.

He's a wonderful lizard. He looks so primitive, almost like some sort of dinosaur. Look at the eyes, set really high up in its head. This is an important survival characteristic for this creature. With eyes set high, it can look up from the opening of its burrow and see any would-be predator moving about, such as a bird of prey or perhaps a snake. And if it feels threatened, it slides down in its tunnel.

What a precious little thing. It's a beautiful animal, just kind of hanging out there. He could be thinking, "There's Jeff Corwin. Do not eat me. I am not edible." Oh, but you're wrong, my spiny-tailed friend. You're quite delicious—so delicious that these lizards are being wiped out in some parts of India. Why does he taste so good? Maybe because he's a vegetarian—90 percent of what this animal eats is plant life.

The *Uromastyx,* a very ancient-looking lizard, but very modern in its ability to survive and take on the challenges put forth by the environment of India. All right, buddy. Back in your hole.

What are you looking at?

I hear you're quite tasty, my spiny-tailed friend.

Maldhari herdsmen also live in Sansan Gir.

And so do their cattle.

Their way of life hasn't changed for centuries.

    Sansan Gir encompasses over ten thousand acres of sun-baked habitat. Besides being home to the last remaining colony of Asiatic lions, Gir is also the traditional home of the herdsmen called the Maldhari. Life hasn't changed much here in the last five hundred years.

Check out this unusual herd.

A lot of water buffalo.

The Maldhari use this land to raise their cattle. As you can see, their herds are made up of very interesting animals, not the typical Brahmin cows that you see in India. These are all water buffalo, and there's a camel over there.

There's even a camel!

There's lots of fodder here for the animals to feed upon. But the risk is that among this high grass, there could be a leopard or a lion. Luckily, if a lion should take your animals—and each year hundreds of domesticated animals are taken by lions and leopards—the government will reimburse you for your loss.

They've invited me to tea.

This is what goes into real chai.

Mmm...delicious!

These guys have asked us to sit down, and they've mixed up a drink called chai. It's the real stuff—made with milk, tea, sugar, and spices—not the Seattle coffeehouse chai. It's delicious. But now I have to say good-bye and go back to work.

See that little nose sticking up from the water? That's a turtle. I'm going after him, but let's just hope a marsh crocodile doesn't come down this waterway. India is an amazing place—here there's a turtle; over there, there could be lion and leopard or crocodile; and there are people here as well. Everybody's living together—there's not a lot of space in India.

Look at that!

There's a nose sticking out of the water!

Hey! No biting, mister.

Got him! And he's biting me. Ow! No biting Jeff Corwin. This is a very interesting turtle, and for some people in India it is a delicacy. This is the flap-shelled turtle or the Indian mud turtle. Why do they call it the flap-shelled turtle? It has a fleshy hinge on the back of its shell.

See the hinge on the back of its shell? That's why it's called a flap-shelled turtle.

These turtles are masters at swimming. Look at the feet. They have dramatic amounts of webbing between the toes, giving this animal the ability to move very quickly through the water.

Here's something else neat about this animal. Look at its face—this turtle has a snorkel. It can submerge its body and be completely invisible from would be predators. Then it can stick its long neck up so the tip of its little snorkel nose just breaks the surface, to take a breath.

All this webbing makes him a fast swimmer.

This snorkel nose lets him breathe when the rest of his body is underwater.

A lot of predators are after these turtles.

These turtles can grow as large as a trash-can lid!

Many creatures prey on these animals, including crocodiles and lions. The flap-shelled turtles eat frogs, invertebrates, fish, and other small animals, as well as plants. These turtles can go up to two years without food, which is amazing. And they can grow as big as a trash-can lid and weigh up to ten or eleven pounds. A really wonderful animal, the Indian mud or flap-shelled turtle. Back to the river you go, my friend.

Isn't this the coolest turtle?

37

Well, here we are. It's been a tremendous, trying journey to get from Ranthambhor to Gir National Park. But this is the moment we've worked so hard for.

We have a good team assembled. Ashwin and Mohammed are park rangers, part of the management team that protects the wildlife of Gir. And we have Dr. Ravi Chellam. He's a wildlife biologist, and he specializes in Asiatic lions—the animals that we hope to find here. Ravi Chellam was one of the first scientists to study the Asiatic lions

We finally made it to Gir National Park.

We're going to see some Asiatic lions!

of Sansan Gir. He's been their outspoken defender for nearly twenty years. He knows their behavior inside and out.

Under normal circumstances, it is absolutely forbidden for visitors to get out of the jeep in this park. But since we're with Ravi and his trackers, we will get an exclusive opportunity to see these great predators from the ground. Hopefully we won't join the cheetal as just another item on the Gir Park menu.

Dr. Chellam knows everything about them.

I hope we don't end up like that.

Ravi tells us that on average lions attack fourteen people each year in the park, and over the years several hundred people have been killed. In the late 1980s and early 1990s, a drought reduced the number of prey animals here, and killings of humans increased. Some twenty-two people were killed in those years, and seven were eaten.

The lions in this park have eaten quite a few people.

A cheetal's alarm call tells us that we're getting close to our goal. What's great about this encounter is that when we think of lions, we usually think of Africa—that's where most lions are found. But

I don't believe it! There's one now.

A lioness resting in the shade.

at one time there were seven subspecies of lions. Today, there are only five left, and this Asian lion one of the rarest and most endangered. In fact, the population of these animals is extremely low. There are only about 324 lions, and they are all here in Gir.

We've heard a male lion calling, and suddenly there's a female in view. And she's not alone—there are two cubs with her.

Lions usually have one to four cubs, but 80 percent of all cubs die within the first two years. The four greatest factors that lead to the death of these lion cubs are disease, starvation, predation, and competition.

This is extraordinary. We're on the ground with the pinnacle predator of Gir, the Asiatic lion.

Isn't she beautiful?

And she has her cubs with her.

The Asiatic lion is truly a noble creature.

The cub is curious, but smart enough to stay hidden.

Not even fifty feet away from us is this glorious lioness, probably weighing 180 to 250 pounds. Her cubs have taken shelter behind her, using their mother as a defense. You would think an animal like this would not have much to defend itself against, but in fact there are leopards and pythons here that prey on young cubs.

It's amazing to see extraordinary predators like these lions just resting with humans so close by. Humans have had an adversarial and competitive relationship with these animals for many, many generations. To be able to sit at the same table without one eating the other is kind of neat.

It appears that this lioness is settling in for a nap. I don't know if I should be offended or flattered—offended that a creature built like Jeff Corwin is no threat to this animal, or flattered because we accomplished our goal. We became a part of the landscape. We were able to creep in close and really experience this splendid beast.

Looks like it's naptime for her and her cubs.

"What an amazing journey! Not only did we see to an Asiatic lion..."

"we also got close to a Bengal tiger."

This was one heck of a journey. We traveled west from Ranthambhor, where we discovered and experienced the Bengal tiger. And we ended up here in Gir, face-to-face with the Asiatic lion. It doesn't get any better than this, so we'll wrap it up. I'll look forward to our next adventure.

Glossary

adversarial in opposition

arid dry

conservationists people who work to protect wildlife and natural resources from loss

diverse varied

domesticated tamed or raised by people

ecosystem a community of living things and the surroundings in which they live

endangered in danger; an endangered species is in danger of dying out

environment the surroundings and conditions that affect living things and their ability to survive

extinction dying out

habitat the place where a plant or animal naturally lives

indigenous growing or living naturally in a particular place

matriarchal led by females

pinnacle top

predation the act of killing and eating animals

predators animals that kill and eat other animals

prop roots a root that serves as a prop or support for a plant

venomous poisonous

Index

Asiatic lion, 25, 26, 38–45

Banyan tree, 12–13
Bengal tiger, 16, 18–19, 21–24
Black-faced langurs, 13–15

Cattle, 31
Cheetal, 18–19, 39–45
Cobras, 6–11
Cubs, 22, 23, 41–42, 44

Deer, 18–19, 20, 24
Defense, 44

Endangered animals, 41
Extinction, 17
Eyes, 15, 28

Feet, 35
Flap-shelled turtle, 34–37
Fur, 20

Hunting, 16

Indian cobra, 6–11
Indian mud turtle, 34–37

Langurs, 13–15
Lion, 25, 26, 38–45
Lizard, 27–29

Maldhari, 30–32
Monkeys, 13–15
Mud turtle, 34–37

Paw, 21
Predator, 21, 25, 27, 28, 35, 39, 44
Pregnancy, 22
Prey, 19
Primate, 13

Sambar, 20, 24
Scat, 20, 27
Serpent, 8, 10
Shell, 34
Snakes, 6–11
Snorkel, 35
Spiny-tailed lizard, 27–29
Spotted deer, 18–19
Swimming, 35

Tatti, 20
Tigers, 16–24
Turtle, 33–37

Vegetarian, 13, 29
Venom, 6, 8

Water buffalo, 31
Webbed feet, 35

48

UNSWORTH ELEMENTARY SCHOOL
5685 Unsworth Road South
Chilliwack, BC V2R 4P5